Computer Science Cooperation Workshop

Wolfram Hardt
Paul Rosenthal
(Hrsg.)

TUDpress

IBS Scientific Workshop Proceedings
Herausgegeben von Stiftung IBS, Wolfram Hardt
Band 1

Computer Science Cooperation Workshop

Wolfram Hardt
Paul Rosenthal
(Hrsg.)

TUDpress

2015

Bibliografische Information der Deutschen Nationalbibliothek
Die Deutsche Nationalbibliothek verzeichnet diese Publikation in der
Deutschen Nationalbibliografie; detaillierte bibliografische Daten sind im
Internet über http://dnb.d-nb.de abrufbar.

Bibliographic information published by the Deutsche Nationalbibliothek
The Deutsche Nationalbibliothek lists this publication in the Deutsche Nationalbibliografie; detailed bibliographic data are available in the Internet
at http://dnb.d-nb.de.

ISBN 978-3-95908-016-3

© 2015 TUDpress
Verlag der Wissenschaften GmbH
Bergstr. 70 | D-01069 Dresden
Tel.: 0351/47 96 97 20 | Fax: 0351/47 96 08 19
http://www.tudpress.de

Alle Rechte vorbehalten. All rights reserved.
Gesetzt von den Herausgebern.
Printed in Germany.

IBS Scientific Workshop Proceedings

Foundation IBS supports the establishment, growth and preservation of scientific oriented interdisciplinary networks. National and international scientists and experts are brought together for research and technology transfer, where the foundation IBS provides the suitable environment for conferences, workshops and seminars. Especially young researchers are encouraged to join all IBS events.

The series *IBS Scientific Workshop Proceedings (IBS-SWP)* publishes peer reviewed papers contributed to IBS - Workshops. IBS-SWP are open access publications, i.e. all volumes are online available at IBS website www.ibs-laubuch.de/ibs-swp and free of charge.

IBS Scientific Workshop Proceedings aim at the ensuring of permanent visibility and access to the research results presented during the events staged by Foundation IBS.

Editorial Board

- Dr. Ariane Heller
  Technische Universität Chemnitz

- Dr. André Meisel
  Technische Universität Chemnitz

- Dr. Olaf Müller
  BMW AG

- Prof. Dr. Wolfram Hardt, Editor-in-Chief
  Technische Universität Chemnitz, Stiftung IBS

www.ibs-laubusch.de/ibs-swp

IBS Scientific Workshop Proceedings

IBS Workshop September 09th, 2013

Table of contents

FPGA-based Image Pre-processing and Overlay Visualization 3
Muhammad Faisal Abbas

Implementing intelligent cars using software components 6
Tomáš Bureš

Secure Vectorization-supported Synchronization and localization in factories 8
Daniel Reißner

Resource-Aware Controller/Scheduler Co-Design 11
Alejandro Masrur

Preface

This volume of *IBS Scientific Workshop Proceedings* collects research work from international cooperation between German and Czech universities. In 2013 the research cooperation was granted an *IBS Scientific Workshop* for the first time to support ongoing scientific exchange and cooperation. This reading on hand, contains a selection of papers, presented in 2013. Key topics are image processing, component based software technologies, synchronization and localization in logistics for factories and resource are controller design.

I am glad to present this first volume of *IBS Scientific Workshop Proceedings* and encourage young researchers to join IBS Workshops.

Prof. Dr. Wolfram Hardt

Chairman Foundation IBS, February 2015

Acknowledgement: We thank DAAD (program Ostpartnerschaften) for supporting this IBS Workshop in September 2013.

FPGA-based Image Pre-processing and Overlay Visualization

Muhammad Faisal Abbas
Technische Universitat Chemnitz
Chemnitz, Germany
abmu@hrz.tu-chemnitz.de

Prof. Dr. W. Hardt
Technische Universitat Chemnitz
Chemnitz, Germany
hardt@cs.tu-chemnitz.de

Dr. Mirko Caspar
Technische Universitat Chemnitz
Chemnitz, Germany
mirko.caspar@informatik.tu-chemnitz.de

Abstract— This paper presents the architecture of Image Pre-processing and Text Overlay Visualization (IPPTOV) system. The designed system is modular and based on Field Programmable Gate Array. It can be used to generate image data for simulation purposes. Two point operations for image pre-processing are facilitated in the system architecture namely, contrast adjustment and brightness enhancement. The proposed system is useful for real-world computer vision systems like error display, statistics display, date time display, subtitling, stock ticker, info messaging, and textual adverts.

Keywords—image pre-processing; overlay visualization; image analysis; FPGA

I. Introduction

With the advent of reconfigurable hardware and their supporting high level design and programming languages, design and implementation of image processing applications in field-programmable gate arrays (FPGAs) have gained tremendous popularity [1]. FPGAs exploit their inherent parallelism and re-configurability to deliver improved performance of image processing techniques [2]. This performance increase and better design of FPGAs is further exploited by deploying an effective image pre-processing unit. Such a unit pre-processes the imaging data before it is fed into the unit concerned with fundamental image processing tasks. Further, such a unit is capable to perform text overlay visualization as a post-image-processing practice. According to Pong P. Chu [3], there are three broad categories of image and text generation and overlay including, bit-mapped scheme, tile-mapped scheme, and object-mapped scheme.

This paper presents architecture of an Image Pre-processing and Text Overlay Visualization (IPPTOV) system. It uses the parallelism of the provided reconfigurable hardware to achieve desired results and performance. By using the external SRAM of the FPGA, this design gets frame by frame pixel data from simulated camera sensor(s) source, buffers it locally, resizes each frame to fit the standard VGA resolution, performs various image filtering methods like contrast adjustments and brightness enhancement, applies a pre-defined textual overlay and transmits the result to a standard VGA display.

The FPGA-based image pre-processing unit is designed to contain sub-modules for handling of a portion of computation on the input imaging data, including contrast adjustment and brightness enhancement at the pixel level.

The remainder of this paper is organized as follows. Section II presents the system architecture and explains the functional design of the system. Section III gives some future research and development directions. Finally, Section IV concludes this paper.

II. Architecture

The underlying modular architecture of IPPTOV is application independent. This allows each key module of the architecture to be designed independently and redundantly, to investigate several design possibilities. For example, image filtering and image displaying modules can communicate with each other without knowing the image creation mechanism. It allows several image generation schemes to be designed and tested with a single image filtering design. For this paper, three different image pre-processing and textual overlay topologies are designed based on device dependent video buffering, without video buffering, and with external video buffering. These filtering topologies observe 1) the impact of image and video buffering, usually an intrinsic requirement for industrial scale image processing systems, and 2) the device's hardware resources, power, and memory utilization when image or video is either not buffered at all or buffered in a limited fashion or with full bandwidth.

Depending on selected image filtering topology and configuration, the system contains six types of modules. The first type of module includes two image creation and pixel generation interfaces. At one time, only one of the image generation interfaces is active. The second type of module is the image pre-processing and filtering interface that supports contrast adjustment and brightness enhancement. The third type of module is the image storage interface. Depending on the selected image filtering topology, this module provides interface to the video memory either in the form of device dependent block RAM or the external memory. The fourth type of module is the overlay visualization control interface, which includes the textual overlay generation and font ROM control. The fifth type of module is the display control. Currently, it supports only standard VGA resolution of 640x480. The application independent design allows the configuration of various high or low resolutions. The sixth type of modules is auxiliary interfaces, which includes a clock divider, a timer, and a digit counter. Fig. 1 shows the architecture of IPPTOV system with the major components.

The system has two modes of operations including, bypass and processing modes. The bypass mode turns off image pre-processing and filtering. In this mode, the original image gets transmitted to display with only textual overlay.

The system has an interface to the external host for communication. This interface is updated based on selected image filtering topology and configuration.

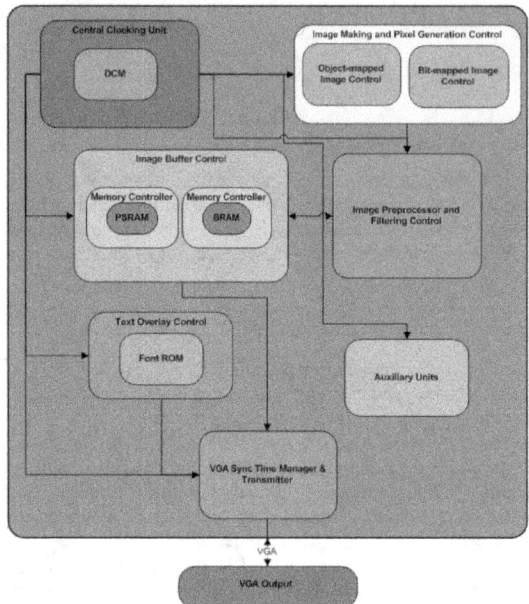

Fig. 1. Architecture of the image pre-processing and text overlay visualization system.

The major components of IPPTOV system are described below.

A. Image acquistion

Three image filtering topologies are used for image generation and processing by using object mapped scheme and bit mapped scheme.

B. Filtering and pre-processing

The point operations are widely used for contrast enhancement, segmentation - color filtering, change detection, masking, and many other applications. This system considers only contrast enhancement and brightness adjustment. From hardware design perspective, these two are implemented using streamed processing mode. We systematically pass each pixel through a single hardware module, which is responsible to perform these functions. Such formation makes it feasible to determine the input ranges at compile time.

C. Overlay visualization

The real-world image processing systems can be made highly informative to the end user via visual overlay, either in text, image, or video format. The proposed system has a textual overlay module to display the coordinates of a moving object, an informative message for the use of IPPTOV system, the "TUC" logo, and the end-of-processing message.

As each of these messages gets displayed on different occasions and different locations, they need to be treated as individual objects. Using the tile-mapped approach, each character is taken as a tile. Thus, a title-mapped character is mapped to a pre-defined pattern instead of representing by a single pixel. The pre-defined pattern is usually defined via a ROM. The proposed design takes a similar approach and stores the 7-bit ASCII code for character patterns as tiles in a ROM. The pattern memory is called font ROM. As the title patterns constitute the font of the character set to be displayed, there are various types of fonts possible.

The 128 (2^7) character patterns are accommodated by a 2^{11}x8 font ROM. In this ROM, the seven MSBs of the 11-bit address are used to identify the character, and the four LSBs of the address are used to identify the row within a character pattern.

D. Storage and buffering

Depending on the filtering topology selection, either no image and video memory is required for pre-processing or an internal block memory or external static memory is utilized for the purpose of storing and buffering image and pixel data.

As object-mapped image creation and pixel generation does not need video memory or buffer, image filtering topology 1 uses no memory for storage of image data. The image data passes directly from image generation stage to filtering and pre-processing phase. In case of filtering topology 2 which uses bit-mapped scheme of image creation and pixel generation, an image or video memory is required. This topology differs from 3 by the size of video buffer. It uses only FPGA device's internal memory available as block RAM.

Filtering topology 3 simulates an external image source and hence needs more storage capacity. The asynchronous static RAM (SRAM) is used for massive storage as a RAM cell is comparatively simpler than a flip-flop cell. However, it has its own challenges as data access is more complex as the system under design is synchronous while SRAM access is asynchronous in nature. Thus, a memory controller is used to interface with the SRAM.

A video data buffer controller using a PSRAM memory for physical storage of imaging data is designed. The memory controller has two roles, storing pixel results and sending the pixel data to any sink system or module which can be a display or any other capable processing node.

E. Display

In order to display the image processing and text overlay visualization results, a VGA-display controller component is designed. A display controller outputs the current pixel coordinates to allow an image source to provide the appropriate pixel values to the video DAC, which in turn drives the VGA monitor's analog inputs. It also provides the sync signals for the VGA monitor.

In order to display an image on the screen, individual pixels are turned on or off on the whole screen. The controller scans the whole screen continuously. The standard VGA controller uses 7 signals in total - two input signals, one input signal from the switch button for resetting of complete display, the other being clock signal via clock-generator module, and five output signals, two data out buses for horizontal and vertical coordinates, and two signals for horizontal and vertical synchronization.

Since the VGA controller requires a pixel clock at the frequency of the VGA mode being implemented, this frequency is generated locally by using a clock down converter which is essentially a clock divider. The generated clock then drives the horizontal and vertical scan counters to produce the correct synchronization signals. In case, the pixel data is coming from the memory controller directly, a pixel data input buffer can be used for display optimization.

F. Auxillary units

Apart from a clock converter needed for generating the correct clock for the display synchronization circuit, the top level design requires two small utility modules including, a counter and a timer unit to facilitate the screen switching and counting of coordinates.

III. FUTURE WORK

Further studies will investigate the implementation of proposed architecture of IPPTOV system. We will investigate the implementation of each major component of the IPPTOV including, the design of image acquisition at the hardware level, hardware description language (HDL) implementation of each filtering topology, the two image filters of pixel brightness and contrast enhancement for pre-processing and filtration subsystem, HDL implementation of video memory and frame buffering approaches, external memory controller interface and the overlay visualization subsystem used for textual overlay at the HDL level, and the HDL design and implementation of the VGA display subsystem.

IV. SUMMARY

This paper provides the functional design and architecture of IPPTOV system. It provides a synopsis of the overlay design and explains the functionality of each module. Further, it elaborates how image is produced via image making module that uses pixel brightness and contrast enhancement component of pre-processing and filtering module. Afterwards, it describes the interfacing between image making module and memory controller and describes the memory controller to store pre-processed or unprocessed imaging data into the external memory. It then, elaborates the working of textual visual overlay that is used to display axes coordinates representing the current location of the moving object in the original image source.

REFERENCES

[1] Y. Said, T. Saidani, F. Smach, M. Atri, and H. Snoussi, "Embedded real-time video processing system on FPGA by image and signal processing," Lecture Notes in Computer Science, 2012, vol. 7340, pp. 85-92.

[2] S. Anitha and V. Radha, "Comparison of image preprocessing techniques for textile texture images," International Journal of Engineering Science and Technology, 2010, vol. 2, pp. 7619-7625.

[3] P.P. Chu, "FPGA prototyping by VHDL examples Xilinx Spartan-3 version," Wiley Interscience, 2008, (chapter 13) pp. 291-319.

Implementing intelligent cars using software components

Tomas Bures, Petr Hnetynka, Jaroslav Keznikl, Frantisek Plasil,
Rima Al Ali, Ilias Gerostathopoulos, Michal Kit
Charles University in Prague, Faculty of Mathematics and Physics
Department of Distributed and Dependable Systems
Malostranské náměstí 25, Prague 1, 118 00, Czech Republic
Email: {bures,hnetynka,keznikl,plasil,alali,iliasg,kit}@d3s.mff.cuni.cz

With the increasing number of mobile computing devices (embedded computers, smartphones,...), there is a growing demand for developing large interconnected systems employing these devices. These systems (usually called Cyber-physical systems — CPS) are typically highly-dynamic, open-ended, robust and self-adaptive. An issue is that currently there are no widely adopted development processes and computational models for CPS.

In our Department of Distributed and Dependable Systems[1] (D3S) we aim at finding an efficient way to design and develop CPS systems; one of our outcomes is a new family of component-based systems called Ensemble-Based Component Systems (EBCS) and specifically the DEECo (Distributed Emergent Ensembles of Components) component system [1] as an instantiation of ECBS. Traditional composition of components is in DEECo substituted via ensembles — each of them being a dynamically formed goal-oriented group of components implicitly exchanging their data. The principles of DEECo and the related development process are below explained via a case-study of electric car sharing (the case-study has been designed and developed in the scope of the EU FP7 project ASCENS[2]).

The case-study deals with efficient sharing of intelligent electric cars (e-cars). Each e-car is assigned a set of points-of-interest (POIs) to be reached within particular time constraints (compiled from the requirements of the users sharing it). The actual schedule has to take into account also the necessity to recharge the vehicle during stop-overs and potential vehicle exchange. For charging, a set of parking-lot/charging-stations (PLCSs) is available. The e-cars have a set of sensors for monitoring battery level and energy consumption and also for monitoring traffic density. To reach its desired POIs, an e-car is expected to communicate with the PLCSs along the route and with other e-cars in vicinity to exchange information about traffic jams, road closings, etc.

Implementing such a system using traditional ways of explicit communication using messages and/or method calls would be difficult since the system architecture is subject to a frequent change (different cars within communication range,

[1]http://d3s.mff.cuni.cz/
[2]http://www.ascens-ist.eu/

closed PLCS, etc.) and has to be decentralized. ECBS (and DEECo in particular) address these challenges by combining several existing approaches and their selected features (depicted in Figure 1).

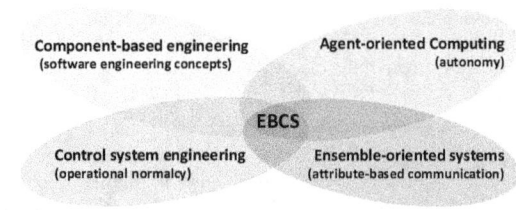

Fig. 1. Approaches combined into EBCS

In DEECo, the e-cars case study is modeled as a set of components and ensembles (each in multiple instances). Each component instance has its knowledge and processes. Knowledge represents the component's state and is organized as a tree data structure. The knowledge is exposed via interfaces, which provide a partial view on the component state. A process is a function operating over the component's knowledge; first, particular input fields of knowledge are atomically retrieved, then the process is executed and finally, outputs are stored again into knowledge fields. Processes run independently (no communication among other processes or components) and they are executed by the run-time framework either periodically or as a reaction to a particular knowledge field change. Interactions among components (knowledge exchange) are determined by definition of ensembles. An ensemble is defined via a membership condition and knowledge exchange function. The membership condition declaratively expresses under which circumstances components form the ensemble (at one moment a single component can be involved in several ensembles — see Figure 2), which then implies the specified knowledge exchange among components in the ensemble. Exchange is performed by the run-time framework. Thus, a component's processes operate solely upon the knowledge local to the component, which is implicitly updated as the component enters/leaves an ensemble.

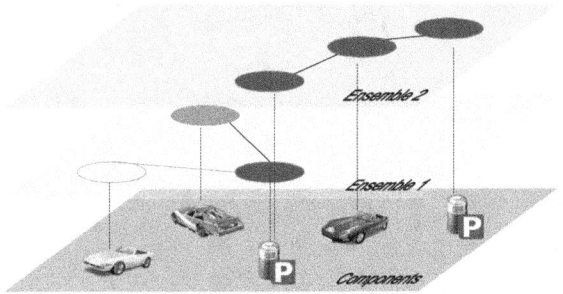

Fig. 2. Components and ensembles

DEECo is equipped with a well-defined computation model allowing timing analysis, with a dedicated high-level design methods called Invariant Refinement Method (IRM) [2] and with a proof-of-the concept implementation called jDEECo[3].

In jDEECo the components and ensembles are realized as plain Java classes with annotations. Knowledge exchange in a distributed environment is implemented using the JavaSpaces middleware (which allows communication via distributed hash-tables). Several case-studies from different application domains (including e-cars) have been implemented in jDEECo. The overall experience shows that by providing a novel software engineering approach for development of highly dynamic, robust, and self-adaptive systems the approach of DEECo is sound and very suitable for CPS.

ACKNOWLEDGMENTS

This work was partially supported by the EU project AS-CENS 257414 and partially received funding from the European Union Seventh Framework Programme FP7-PEOPLE-2010-ITN under grant agreement n°264840.

REFERENCES

[1] T. Bures, I. Gerostathopoulos, P. Hnetynka, J. Keznikl, M. Kit, and F. Plasil, "DEECO: an ensemble-based component system," in *Proceedings of CBSE 2013, Vancouver, Canada*, June 2013, pp. 81–90.
[2] J. Keznikl, T. Bures, F. Plasil, I. Gerostathopoulos, P. Hnetynka, and N. Hoch, "Design of ensemble-based component systems by invariant refinement," in *Proceedings of CBSE 2013, Vancouver, Canada*, June 2013, pp. 91–100.

[3]https://github.com/d3scomp/JDEECo

Secure Vectorization-supported Synchronization and localization in factories

Daniel Reißner, Wolfram Hardt, Technische Universität Chemnitz, Germany

Abstract— for exact energy-efficient production of goods in factories, tasks have to be synchronized. For synchronization, rate of processes must be known, while simple detection on nodes is often not enough. For example detection of carrier of super positioned processes could not ensure that single carrier of one process have to skip one station, as well as uncalculated parameters like relocation and transport duration of mobile waggons are not considered. In order to deal with this and further unknown values, tracking of process in process dynamics is required. In this paper a vectorization-based approach for tracking of process with mobile pathfinder nodes is introduced. Furthermore resulting Energy savings are demonstrated on self-organized application gradient routing. By integration of the application gradient preference and the more exact vector-approach resulted timing in ciphering key generation, an energy-efficient approach for safe data transfer in sensor networks is presented.

Keywords— synchronization, localization, adaptive, robust, preference-based, energy-efficient, routing

I. INTRODUCTION

In order to organize maintenance and repair processes in factories various systemized approaches exist [1]. In order to allow optimization under complex dependencies between approaches and unknown influences, an adaptive mechanism for observation of improvement of tested proceeding at runtime was proposed [2]. Thereby complex nonlinear behavior of environment is parameterized, compressed and evaluated in node preference [3].

Existing algorithms for localization and mapping with landmarks are Kalman filter based and require for knowledge about distribution e. g. mean, variance of application area, possibility to assume Gaussian distribution FastSLAM, e.g. for calculation of covariance matrix [4].

Untrained parameterized feature detector tests with FAST showed noisy behavior for movement related changing images in video data.

Following a training free image processing pipeline using vectorization is introduced for reliable detection of antenna landmarks in factory for improving process flow control of semi-finished good and sensor-actor-task data routing in factories.

II. VECTORIZATION ALGORITHM

In order to track sensor nodes, characteristic properties are antenna and barcode combination as shown in Fig. 1. In order to detect sensor nodes antenna most significant identified property was straight lines in canny edge detected image. In order to evaluate canny result a vectorization contour detection was applied to canny image. The resulting contours are image depending the requested antenna and sometimes compounds with other contours. In order to more reliable detect the antenna the bounding box of compound is evaluated against adaptive antenna height and aspect ratio. This results in reliable detection of antenna as shown in Fig 1. By using the bounding box the algorithm is very tolerant again variations of canny threshold, because most significant property seems to be also detectable in noisy results, also if it results in a strongly deformed contour. In order to distinguish from other vertical structures in factory or classify other structures second integrated parameter is a test for barcode/QR-code. This way only sensor antenna with relative proportion to barcode could be filter out of results. In factory under controlled conditions it is possible to further simplify detection by mounting panel to ensure that there are no disturbances of background.

The antenna detection is invariant of the horizontal rotation view angle of camera reliable detectable in case of clear view to antenna. This way as shown in Fig. 1 it is possible to use the algorithm to detect antenna from top if rotated horizontal upper image and from the side in case of antenna in vertical position.

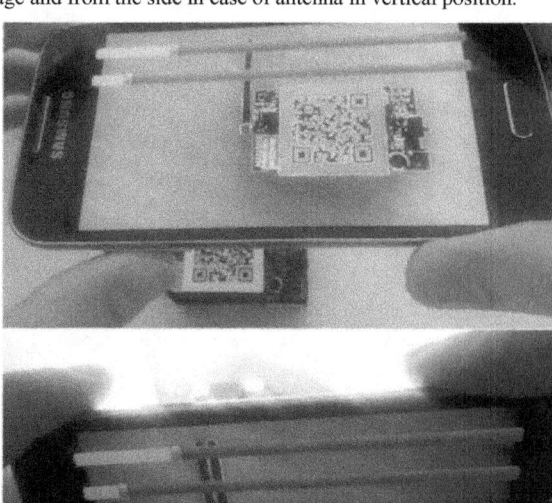

Fig. 1: antenna detection (red rectangle) with uncalibrated canny threshold (in wide ranges tolerant) and calibrated minimal height and aspect to exclude wrong rectangles. Usable to detect barcodes and check if detection position a

machine or Microcontroller for exact rate determination of nodes. (sensor nodes placeable in calibrated situation, wall mountable), also on pc height adaptive version tested prefereing aspect and adapt to best height on base of heighest heigh in image.

At the beginning of pipeline before canny edge detection camera received image was rescaled to low resolution image, in order to faster detect on smartphone. Because of significant lower processing requirements iterative resolution increase and test of algorithm seems appropriate. The whole pipeline can be seen as a control loop adjusting binary search like canny threshold, image height and resolution until antenna is detected under height, aspect, relative barcode conditions for more general solution. Compared to pixel intense circle detection with Hough circle breaking frame rate to insufficient 3 fps, in case of our higher level simple vectorization shape detection, real time detection is possible allowing detection in case of moved image on carrier belt.

In order to generalize further for moved vehicle a model of god and bad detected rectangle landmarks is introduced. Thereby while moving towards or away from a landmark height is adapted to keep tracking. Based on height related distance to landmark in case of more than one landmark position estimation is possible. By integrating position with error surrounding vector to objects and vector to target superposition error surrounding paths to target are construct able at each point in time.

III. APPLICATION GRADIENT ROUTING IMPROVEMENTS

By integration of localization information it is more accurate possible to detect travel time between static sensor nodes on carrier belt considering process of current node. Therefore static camera or camera on mobile pathfinder counts time between process relevant node detection under process conditions. This way processing time from node to node is available. Furthermore by image base pixel counting on horizontal row along integrated moved images in relation to known height, pixel accurate distance measurement from sensor node to sensor node, relevant in process is possible. Accurate timing and localization information on application area allows synchronizing sleep wake periods of sensor nodes to processing times.

In order to reduce energy requirements further from process information and sending ranges energy zones are able to establish. This way as shown in Fig. 2 only gradient is propagated inside of one zone and an alias gradient is propagated as representant, to transfer in other zone with many other gradient values. In Fig. 2 node 1 wants to send to node 19 which is higher than limit 16 and smaller than next limit 32 so it sends to alias 16. On relay node border, request to 19 could by value dereferenced and transfer from 2 over 18 to 19 is done. In special case of mac-id equal to sensor-actor-task application functionality, appgradient routing realizes routing to mac addresses. In case of no preference parameters like energy and only gradient indicator for neighbour, application gradient routes like by using of simple routing table.

Compared to static subnet masks, energy zones are more adapted to requirements of factory. Therefore a distribution mechanism for appgradient id is introduced. In case a node receives a not a-prior-knowledge fitting gateway id, it will not transfer gradient information in initial or on request phase. Instead it will listen until the gradient is propagated through the network and at the end selects the next free gradient value and propagates it. In case of other also listening for new id nodes winner is the node which first registers on server. For long time integrated nodes one time confirmation by server for right appid is appropriate. For short time applying moving nodes a neighbour and past aware free gradient assignment from neighbour for direct communication may be enough for certain mobile node to energy group and energy group to mobile node communication. Also in case of mobile nodes, differentiation by energy-group-node to which mobile node applied is sufficient. In case of two nodes apply at same time at same energy-group-node, energy-group-node can handle id distribution.

By energy groups the communication effort may be evaluated as low enough for continuous gradient update over some time to identify disturbance behaviour relevant shorter paths. Application gradient forms a mesh of order relations usable for logical location network forming detailed by timely detections. Timely synchronization of inner nodes to application area related sending of leaf nodes is possible by introduction of delta values in preference model. This way node counts time to next detection managed by preference model deltas and this way can sleep in case of time to next node receive bigger then rentable sleep time.

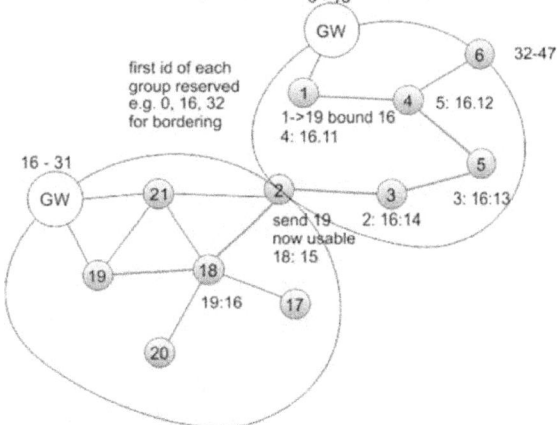

Fig. 2: Gradient information realizes mesh structure usable for localization, enhanced resolution id/mac routing with factory related zones, overload id as appid possible, no routing table update after complete propagation, no routing table but pointered structure for synchronisation + res decision structuring -> pref model (more general number based selection principle)

IV. RESULTS

In order to exploit the higher accuracy of tracking scenario was realized in omnett++. Thereby synchronization to application area processes allowed for identification of sleep periods. Furthermore inner nodes sometimes identified sleep periods from tracking the send periods using delta information in preference model. Simulation results for a scenario with moving carriers with and without the sleep periods are illustrated in Fig. 3. By introduction of sleep periods energy is saved linear in the sleep duration of nodes per process iteration for same process superposition scenario for each iteration.

Image based antenna detection was realized using nvidia tegra development pack for android. Barcode detection was realized using the open ZXing library and related barcode scanner. Detection of antenna was on top as well as on side robust to moving image possible as shown in Fig. 1.

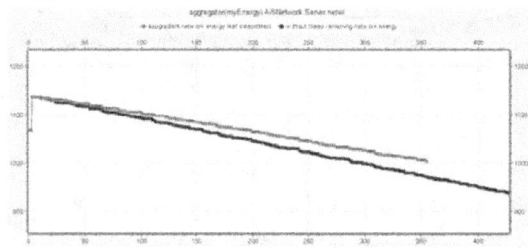

Fig. 3: Using sleep periods, energy is saved linear in transferred data

V. CONCLUSSION AND OUTLOOK

In this paper a detection strategy for sensor nodes based on vectorization was presented which improves energy efficiency of routing by allowing more accurate sleep times.

Canny detection is further improved by more contrast antennas. This way by introducing antennas in light colour in factory could improve detection accuracy. Tracking of height of aspect rectangles could be transferred to further compound horizontal and vertical or rotated structure line detections. Nearer object could be detected by fast moving structures as introduced by chain elongation detection. Further data could be fusioned by selection or averaging best parameter integrating gps, laser distance, radar, ultra sonic, infra read location information. Finally by adaptive zooming image areas disturbed image segments could be regenerated in order to detect better verctorization shapes.

REFERENCES

[1] A.V. Kizim, Establishing the Maintenance and Repair Body of Knowledge: Comprehensive Approach to Ensuring Equipment Maintenance and Repair Organization Efficiency Volgograd State Technical University, Volgograd, Russian Federation, JCKBSE 2014

[2] M. A. Andreevich, ANALYSIS OF THE EXISTING ALGORITHMS FOR THE NAVIGATIONAL AND TERRAIN MAPPING SYSTEMS APPLICABLE IN MOBILE ROBOTIC PLATFORMS. Volgograd State Technical University, Volgograd, Russia, JCKBSE2014

[3] Watteyne, T.; Pister, K.; Barthel, D.; Dohler, M.; Auge-Blum, I., "Implementation of Gradient Routing in Wireless Sensor Networks," *Global Telecommunications Conference, 2009. GLOBECOM 2009. IEEE*, vol., no., pp.1,6, Nov. 30 2009-Dec. 4 2009

[4] Buchhorsts, C.; Dujovne, D., "Natural Gradient Routing: Sink convergence using data as guide," *Communications (LATINCOM), 2013 IEEE Latin-America Conference on*, vol., no., pp.1,6, 24-26 Nov. 201

Resource-Aware Controller/Scheduler Co-Design

Alejandro Masrur
Department of Computer Science
TU Chemnitz, Germany

Abstract—Embedded systems usually implement a set of control loops to interact with their environment. These control loops are associated with quality-of-control (QoC) requirements, which are necessary to achieve the desired behavior. For example, control loops might be required not to exceed a specified settling time, or to have a given stability margin, etc. However, since most embedded systems are implemented upon limited resources, it is sometimes difficult to reliably guarantee all QoC requirements. In particular, since delay might affect control performance, the effect of scheduling on control algorithms needs to be considered. In this paper, we discuss some techniques for controller/scheduler co-design that allow efficiently utilizing available resources and, at the same time, meeting all QoC requirements. In particular, we analyze scheduling strategies for control messages on a mixed time-/event-triggered bus such FlexRay in the automotive domain.

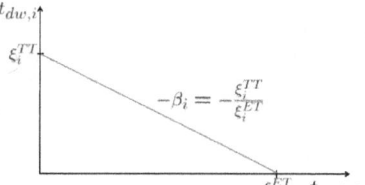

Fig. 1. Relation between $t_{dw,i}$ and $t_{wait,i}$

I. INTRODUCTION

We are concerned with distributed architectures where multiple control applications are executed. In particular, we consider that control applications are mapped onto spatially distributed electronic control units (ECUs) communicating via a hybrid communication protocol such as FlexRay [1]. we study how to better design control applications taking architectural properties into account. More specifically, we address the problem of scheduling control messages on the bus such that communication delay can be reduced and, hence, a better performance can be achieved. At the same time, resource usage should be optimized.

FlexRay allows a *zero/negligible* communication delay when all control messages are mapped onto the static segment of the bus with perfectly synchronized time-triggered (TT) slots and ECUs. Clearly, the controller based on such zero-delay communication leads to a good control performance. However, such TT implementations might be overly expensive because of their high communication bandwidth requirements.

On the other hand, if control messages are mapped to FlexRay's dynamic segment, the resulting implementation suffers from the usual temporal non-determinism, i.e., the communication delay varies with the *priority* and the current scheduling situation on the bus. In such an event-triggered (ET) scheme, a controller is designed based on the *worst-case delay* and might results in a poor control performance.

In this paper we investigate an intermediate possibility where the aim is to achieve control performance close to a purely TT implementation, but using *fewer* TT slots than what would be necessary for purely TT communication. Towards this, we exploit the fact that the time required by a control application to reject an external disturbance (or *response time*) is considerably lesser with the controller based on TT communication compared to the one based on ET communication. To meet a specified response time requirement of a control application, we appropriately switch between the TT and ET modes as originally proposed in [2].

The paper in hand is based on previous work [3] and introduces a *schedulability analysis*, since the number of allocated TT slots is less than what is required for *all* control messages to be accommodated. Hence, in the event of a disturbance, an application might have to wait (depending on whether its associated TT slot is occupied or not) before it may switch from an ET to a TT mode. Designing such a *control performance-oriented scheduling* is the topic of this paper.

In particular, if a controller is switched from ET to TT communication, it is necessary to determine how long this is going to *dwell* in a TT scheme so as to achieve the desired behavior – controllers are switched only once to TT where they remain until the system has settled. This dwell time is a function of the waiting time that a controller spends in ET before switching to TT. In general, the longer this waiting time, the shorter the dwell time is going to be. This is shown in Fig. 1. This needs to be considered when analyzing schedulability of the system.

II. PROBLEM FORMULATION

We consider a set of multiple control applications C_i with sampling period p_i ($i \in \{1, 2 \ldots n\}$) that run on a distributed architecture of the form shown in Fig. 2. The control applications are represented by:

$$x[k+1] = A_i x[k] + B_i u[k], \qquad (1)$$

where $x[k]$ is the plant state, $u[k]$ is the control input and A_i, B_i are the system matrices of C_i. Each C_i is composed of three tasks $T_{s,i}$ (measures $x[k]$), $T_{c,i}$ (computes $u[k]$) and $T_{a,i}$ (applies $u[k]$ to the actuator/plant)—see (1). Such tasks are then mapped onto distributed ECUs which are connected via a hybrid communication bus as FlexRay.

As mentioned above, each communication cycle on the bus is divided into time-triggered (or static) and event-triggered (or dynamic) segments. On the TT segment, the tasks are given access to the bus (or allowed to send messages) only at their predefined *slots*. On the other hand, the tasks are assigned *priorities* in order to arbitrate for the access to the ET segment. Further, we consider the following distributed setup:

- The tasks $T_{s,i}$ and $T_{c,i}$ are mapped onto the same ECU which is attached to the corresponding sensors.
- The tasks $T_{s,i}$ and $T_{a,i}$ that belong to a particular control application are triggered *periodically* with the same period (which is dictated by the sampling time p_i). The

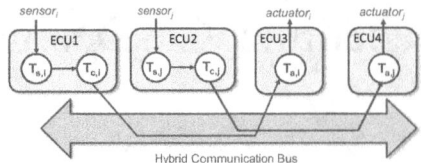

Fig. 2. The distributed cyber-physical architecture in this paper

triggering of $T_{s,i}$ and $T_{a,i}$ is synchronized with a given slot on the static segment of the bus.
- The execution times of $T_{s,i}$, $T_{c,i}$ and $T_{a,i}$ (in the order of a few μs) are negligible compared to the sampling period p_i (in the order of tens of ms).
- Every controller task $T_{c,i}$ can send messages (to $T_{a,i}$) either over the static or the dynamic segment of the bus.

III. SLOT SHARING AND SCHEDULABILITY

To determine the number of necessary slots on the static segment, we first need to decide on how control applications will access slots. In this paper, we focus on a partitioned scheme where each application is assigned to a single slot such that it always uses the same slot when transmitting over the static segment. To determine the necessary number of slots, we need to analyze the schedulability of a set of applications on one slot. Based on such a schedulability analysis, we can allocate applications to one or more slots accordingly.

A. Schedulability Analysis

All C_i sharing one slot on the static segment are assigned priorities according to their criticality. For this purpose, we make use of the Deadline Monotonic (DM) policy, i.e., the shorter the deadline of a C_i, the higher its priority on the given slot.

From our previous discussion, we know that a control application can be switched at most once between the ET and TT regime during a disturbance. Otherwise, the stability of the switching would be compromised. That is, once an application C_i has access to a TT slot, it requires blocking the slot for $t_{dw,i}$ amount of time (i.e., until it finishes transmitting $\frac{t_{dw,i}}{p_i}$ messages). As a result, the scheduling of a sequence of messages on the static segment must be implemented in a non-preemptive manner.

Independent of its priority, an application C_i will have to wait to have access to the TT slot, if this is being used by another application. This increases its waiting time $t_{wait,i}$ of C_i. Hence, its demand for zero-delay communication $t_{dw,i}$ decreases as shown in Fig. 1. However, the overall response time of the application increases with $t_{wait,i}$. This is because the parameter $\beta_i = \frac{\xi_i^{TT}}{\xi_i^{ET}}$ is always less than one.

The schedulability of a control application C_i on a shared TT slot will then be guaranteed, if the following condition holds for every possible $t_{wait,i}$: $\xi_i^d \geq (1 - \beta_i) t_{wait,i}$.

Hence, to test the schedulability of C_i, we need to find the greatest possible $t_{wait,i}$ (denoted by $\hat{t}_{wait,i}$) which leads to the worst-case response time of C_i (denoted by $\hat{\xi}_i$). This occurs when C_i suffers the maximum possible interference due to higher-priority applications. For this, we will consider that all higher-priority applications C_j interfering with C_i require their maximum possible transmission time on the shared slot, i.e., $t_{dw,j} = \xi_j^{TT}$. This assumption is pessimistic since $t_{dw,j}$ actually decreases with the blocking time suffered by C_j.

However, this allows us to simplify the analysis and leads to a *safe* schedulability condition. Under this assumption, the worst-case interference on C_i clearly occurs when it needs to have access to the TT slot together with all higher-priority C_j (sharing the same slot). This again happens when all higher-priority C_j and C_i undergo disturbances at the same time – assuming an inter-arrival of disturbances r_i. Since the scheduling is non-preemptive, there will be blocking time due lower-priority application.

Computing $\hat{t}_{wait,i}$ and $\hat{\xi}_i$ here has some similarities with computing the worst-case response time in a fixed-priority non-preemptive scheduling like the one of CAN [4], [5]. That is, we need to compute the response times of all *jobs* of that task within its *maximum busy period* [5].

In our case, the task is given by a control application C_i sending a certain number of consecutive messages over a shared slot. The maximum busy period $t_{max,i}$ of a C_i is then the largest time interval in which the shared slot is constantly being used by higher-priority control applications and by C_i itself. For ease of exposition, we assume that $t_{max,i} \leq r_i$ holds for all C_i in this paper, i.e., there is only one transmission of $\frac{t_{dw,i}}{p_i}$ messages of C_i within its busy period $t_{max,i}$. This way, we only need to compute the response time ξ_i of the sole *job* of C_i within $t_{max,i}$ to obtain its worst-case response time $\hat{\xi}_i$, which can be done in the following manner:

$$\xi_i = \xi_i^{TT} + (1 - \beta_i) b_i + (1 - \beta_i) \sum_{j=1}^{i-1} \left\lceil \frac{\xi_i}{r_j} \right\rceil \xi_j^{TT}, \quad (2)$$

where $b_i = \max_{k=i+1}^{n}(\xi_k^{TT})$ denotes the maximum possible blocking time due to lower-priority applications suffered by C_i and n is the number of applications. Without loss of generality, we assume in Eq. (2) and in the remainder of the paper that applications are sorted in order of decreasing priority (i.e., C_j has higher priority than C_i and C_i has higher priority than C_k for $1 \leq j < i < k \leq n$). Clearly, if ξ_i exceeds ξ_i^d, C_i is not schedulable on the shared slot. On the other hand, if there is a convergence value prior to ξ_i^d, then C_i can meet its deadline and is schedulable.

IV. CONCLUDING REMARKS

In this paper we proposed a switching strategy for distributed control applications communicating via a hybrid event-/time-triggered protocol. The response times of the control applications are considerably shorter in the TT compared to the ET mode. However, a TT implementation essentially results in poor bus utilization and hence in an expensive design. The approach in this paper allows for a performance close to that of a purely TT scheme using fewer TT slots.

REFERENCES

[1] "The FlexRay Communications System Specifications," Ver. 2.1, www.flexray.com.
[2] D. Goswami, R. Schneider, and S. Chakraborty, "Re-engineering cyber-physical control applications for hybrid communication protocols," in *Design, Automation and Test in Europe (DATE)*, Grenoble, France, 2011.
[3] A. Masrur, D. Goswami, S. Chakraborty, J.-J. Chen, A. Annaswamy, and A. Banerjee, "Timing analysis of cyber-physical applications for hybrid communication protocols," in *Proceedings of the Conference on Design, Automation and Test in Europe (DATE)*, Dresden, Germany, 2012.
[4] K. Tindell, H. Hansson, and A. Wellings, "Analysing real-time communications: Controller area network (CAN)," in *Real-Time Systems Symposium (RTSS)*, 1994.
[5] R. Davis, A. Burns, R. Bril, and J. Lukkien, "Controller area network (CAN) schedulability analysis: Refuted, revisited and revised," *Real-Time Systems*, vol. 35, no. 3, pp. 239–272, 2007.